The Sun

The Source of All things

JD ARDEN

The Sun

Preface: A Fiery Beginning

The Sun is many things to us. It is a blazing sphere of hydrogen and helium, a relentless nuclear furnace that has burned for over 4.6 billion years. But it is also the warm embrace of a summer's day, the orchestrator of life on Earth, and a timeless muse for poets, artists, and dreamers. For as long as humans have gazed at the sky, the Sun has been both a source of wonder and an enigmatic force of nature—a symbol of constancy and a reminder of the immense power that dwarfs all earthly concerns.

Yet, the Sun is not just a distant star. It is the unyielding center of our existence, shaping everything from our seasons to our sense of time. Its light fuels the photosynthesis that feeds the world, while its heat sustains the delicate balance of Earth's ecosystems. And when it erupts in fiery outbursts of solar flares, it sends ripples through our technology-dependent lives, disrupting satellites and power grids.

In this book, we'll move beyond what you learned in school about the Sun. We'll uncover its hidden complexities, from the strange patterns of its magnetic fields to the silent yet dramatic storms raging across its surface. We'll explore how the Sun has shaped the evolution of life, as well as its influence on cultures, myths, and the very fabric of human thought. And finally, we'll peer into the distant future, contemplating the star's eventual demise and what that means for our solar system—and perhaps our species.

The Sun is not just a ball of fire in the sky. It is a teacher, a tyrant, and a companion. This is its story—and ours.

Chapter 1: The Sun We Know

The Sun is the ultimate paradox. It is both ancient and immediate, a burning cauldron of nuclear fire that shapes every moment of our existence. For all its brilliance, however, the Sun remains a distant mystery, hiding its secrets behind an aura of familiarity. We greet it each morning, we measure our lives by its rise and fall, yet we rarely consider its true nature. To know the Sun is to know the force that defines us—and to confront the fragility of the life it sustains.

At its heart, the Sun is a factory of unimaginable energy. Deep within its core, hydrogen atoms collide with such intensity that they fuse into helium, a process that releases vast amounts of energy. This nuclear fusion powers the Sun, creating light that takes thousands of years to escape its dense interior. By the time this light reaches Earth, it has traveled an incredible journey through time and space. The sunlight warming your face today began its story before humanity even existed.

But the Sun is not a simple ball of fire. It is plasma, a state of matter where atoms lose their electrons and create a charged, churning sea. This plasma is not static. It flows and shifts in immense patterns, driven by magnetic fields that defy logic and order. The Sun is alive with movement: bubbling convection cells the size of continents, towering loops of plasma following magnetic lines, and sudden eruptions of energy that ripple across the solar system.

If the Sun were a painting, it would be one of chaos and contradiction. Its photosphere—the "surface" that we see—is deceptively calm at first glance, a glowing disk that bathes Earth in a steady light. But this serene appearance is an illusion. Up close, the photosphere is a battlefield of granules and sunspots, regions of violent activity shaped by the Sun's magnetic forces. Sunspots, cooler and darker patches on the surface, seem almost gentle compared to the solar flares that sometimes erupt from them, sending waves of charged particles hurtling into space.

It is these flares and coronal mass ejections that remind us of the Sun's power. We think of it as a distant benefactor, but it is also a force of destruction. Solar storms can disrupt satellites, interfere with power grids, and create breathtaking auroras. These storms are the Sun's way of reminding us that, for all our technological advancements, we remain tethered to its whims.

The Sun is not static; it has a rhythm. Every 11 years, it undergoes a cycle of activity, a kind of solar "heartbeat." During the solar maximum, sunspots multiply, and flares become more frequent. Magnetic fields twist and tangle, flipping polarity at the height of this cycle before the Sun settles into relative calm. These cycles are not perfectly predictable, and they occasionally deviate in ways that have profound consequences. During the Maunder Minimum of the 17th century, the Sun entered an extended quiet phase that coincided with the Little Ice Age, a period of cooler temperatures on Earth.

This rhythm of activity and calm raises questions about the Sun's role in shaping Earth's climate. How much of what we experience as weather, or even climate change, can be traced back to the subtle influences of solar energy? Scientists continue to explore these connections, but the answers remain elusive.

For all we know about the Sun, there is so much that we don't. The corona, the Sun's outer atmosphere, is one of its greatest mysteries. While the photosphere is about 5,500 degrees Celsius, the corona blazes at millions of degrees—an inexplicable reversal of expected physics. How does the Sun's energy leapfrog into its outermost layer, defying the logic of thermodynamics? The answer lies somewhere in the Sun's magnetic complexity, but even the most advanced solar observatories have yet to fully decode the mystery.

The Sun also sings. Vibrations ripple through its plasma, creating oscillations that resonate like the tones of a musical instrument. These "solar songs" are inaudible to the human ear but provide scientists with valuable clues about the Sun's internal structure. By studying these vibrations, researchers have mapped the Sun's core, much as seismologists use earthquakes to understand the Earth's interior.

And then there is the Sun's future—a story both comforting and unsettling. For now, the Sun is stable, burning its fuel at a rate that ensures billions more years of life. But it is a finite star. In about five billion years, it will swell into a red giant, consuming Mercury and Venus and potentially engulfing Earth. Eventually, it will shed its outer layers,

The Sun

leaving behind a white dwarf—a tiny remnant of the giant it once was. This quiet end is the fate of most stars, a humbling reminder of the universe's cycles of birth and decay.

Despite its immensity, the Sun is losing weight. Every second, it converts four million tons of its mass into energy, radiating it into space. Over time, this loss will alter its gravitational pull, subtly shifting the orbits of the planets. The changes are so gradual that we won't notice them in our lifetimes, but they are a reminder that even the Sun is not eternal.

The Sun is more than a star. It is a paradox—a force that gives life but holds the power to destroy it. It is constant yet ever-changing, familiar yet full of secrets. To look at the Sun is to see not just a ball of light, but a story of complexity, chaos, and cosmic significance.

The Sun we know is only a fraction of its true self. What lies beneath its surface, what it has yet to teach us, and what it means to us as a species—all of this remains part of its unfolding mystery. As we continue to learn about this extraordinary star, we are also learning about the forces that shape our lives and the universe itself.

The Sun is not just our star. It is our beginning, our present, and, in some distant future, it will be our end.

Chapter 2: Cycles of Fire

The Sun, for all its grandeur, is not a constant entity. It breathes, it pulses, it changes. To the naked eye, it may seem like an eternal, unchanging presence in the sky, but this illusion masks a complex reality. The Sun lives by cycles—vast, intricate rhythms that echo across the solar system. These cycles govern its energy, its behavior, and even its relationship with Earth. They are the heartbeats of a star, and they carry with them both creation and destruction.

The most familiar of these cycles is the solar activity cycle, an 11-year pattern of magnetic upheaval that transforms the Sun from a serene, steady star into a turbulent cauldron of storms. During the solar minimum, the Sun's surface is deceptively calm, with few visible sunspots. But as the cycle progresses toward the solar maximum, the Sun erupts with activity. Dark sunspots dot its surface like scars, while powerful solar flares and coronal mass ejections hurl energy and particles into space. These moments of fury are not mere spectacles—they ripple through the fabric of the solar system, shaping the fates of planets and moons.

Sunspots themselves are enigmatic features, cooler and darker regions on the Sun's photosphere. They are created by intense magnetic fields that emerge from the Sun's interior, tangling and twisting as they break through the surface. These magnetic fields trap plasma, reducing the flow of energy and creating the cooler, darker patches we see. But sunspots are more than simple blemishes; they are harbingers of greater forces at play.

Solar flares, the sudden, explosive releases of energy from the Sun's atmosphere, often emerge near sunspots. These flares can release as much energy as billions of nuclear bombs in a matter of minutes. They send X-rays and ultraviolet radiation streaming into space, disrupting communication systems and satellite operations if they collide with Earth's magnetic field. Even more dramatic are coronal mass ejections (CMEs), massive bubbles of charged particles ejected from the Sun's corona. When these particles reach Earth, they can create geomagnetic storms that light up the skies with auroras while wreaking havoc on power grids and electronic systems.

These cycles of solar activity are not mere curiosities for scientists; they have real consequences for life on Earth. The Carrington Event of 1859, the most powerful geomagnetic storm on record, disrupted telegraph systems worldwide and created auroras so bright they were visible in tropical latitudes. If an event of similar magnitude were to occur today, the impact on our technology-dependent world would be catastrophic, potentially knocking out satellites, GPS systems, and power grids for months.

What drives this 11-year cycle is a magnetic dance that takes place deep within the Sun. The Sun's magnetic field is generated by a process called the solar dynamo, a complex interaction between its plasma and its rotation. As the Sun rotates, its equator moves faster than its poles, stretching and twisting magnetic field lines. Over time, these field lines become so twisted that they snap, realign, and ultimately reverse polarity. This magnetic flip is the climax of the solar cycle, a moment when the Sun's energy and activity reach their peak.

But the 11-year cycle is not the only rhythm by which the Sun lives. There are longer cycles—decades and even centuries long—that influence the Sun's behavior in subtler ways. The Maunder Minimum, a period from roughly 1645 to 1715, saw an almost total absence of sunspots, coinciding with a "Little Ice Age" of cooler temperatures on Earth. Scientists are still unraveling the connections between these extended solar quiet periods and Earth's climate, but the implications are profound.

There is also evidence of grand solar cycles spanning thousands of years, patterns that we are only beginning to understand. These cycles may hold the key to understanding not just the Sun's behavior, but also its impact on the history of our planet. Did prolonged periods of high or low solar activity play a role in shaping ancient civilizations, influencing weather patterns, or even triggering societal collapses?

The Sun's cycles are a reminder that its constancy is an illusion. What we perceive as steady and reliable is, in truth, an intricate dance of forces beyond our control. This dynamism is not just a feature of the Sun; it is a fundamental aspect of stars. Across the universe, stars pulse and breathe

The Sun

in ways that mirror the rhythms of life itself. They are born, they grow, they change, and they die.

What makes the Sun's cycles particularly fascinating is their impact on us. The Sun is not merely a distant star; it is an active participant in the story of Earth. Its cycles shape our weather, influence our technology, and perhaps even touch the patterns of our thoughts. We are creatures of the Sun, bound to its rhythms in ways we are only beginning to understand.

There is a certain humility in acknowledging this connection. For all our technological advancements, we remain at the mercy of a star's magnetic whims. And yet, there is also a sense of wonder. To study the Sun's cycles is to glimpse the heartbeat of a universe in motion, to see in its rhythms a reflection of the cycles that govern all life. The Sun is not just the center of our solar system; it is the pulse of our existence.

In its cycles of fire, the Sun reveals itself as more than a source of light and heat. It is a force of creation and destruction, a constant presence that is anything but static. As we uncover the secrets of these cycles, we are reminded of the profound interplay between chaos and order that defines not just the Sun, but the universe itself.

The Sun

Chapter 4: Storms from the Sun

If the Sun's magnetic field is its heart, then its storms are its unrestrained outbursts—chaotic, unpredictable, and profoundly impactful. The serene light that graces our skies every day belies a tempestuous nature that can lash out with unimaginable power. These solar storms, violent expressions of energy and plasma, ripple across the solar system, influencing not only the planets but also the very fabric of human activity.

A solar storm begins with the Sun's magnetic tension. Deep within its core, magnetic field lines twist and contort under the churning forces of plasma. Sometimes, these tangled lines snap and reconnect, releasing energy in a sudden, cataclysmic burst. This event, known as magnetic reconnection, is the genesis of a solar flare—a flash of light and energy that radiates outward in X-rays and ultraviolet waves.

Solar flares are extraordinary in their intensity. A single flare can release as much energy as billions of nuclear bombs. They occur in areas of heightened magnetic activity, often near sunspots, and they can erupt with little warning. From Earth, they appear as brilliant flashes of light on the Sun's surface, but their effects are far-reaching. When the radiation from a solar flare reaches Earth, it can ionize the upper atmosphere, disrupting radio communications and GPS signals.

Even more dramatic than flares are coronal mass ejections, or CMEs. While a solar flare is primarily a burst of radiation, a CME involves the ejection of massive amounts of charged particles and plasma from the Sun's corona. These particles, traveling at speeds of up to 3,000 kilometers per second, form a cloud that barrels through the solar system. If Earth lies in their path, the consequences can be profound.

When a CME collides with Earth's magnetic field, it creates a geomagnetic storm—a disturbance that can disrupt power grids, interfere with satellite operations, and even affect the migration patterns of animals that rely on Earth's magnetic field for navigation. The charged particles interact with the atmosphere, producing spectacular auroras that light up the skies at high latitudes. These auroras, shimmering curtains of green, pink, and violet, are perhaps the most beautiful

The Sun

reminders of the Sun's power, but they are also a testament to its potential for disruption.

The Carrington Event of 1859 is a stark example of the Sun's capacity for chaos. This geomagnetic storm, the most powerful ever recorded, disrupted telegraph systems across the globe, with some telegraph lines sparking and catching fire. If a similar event were to occur today, the impact on our technology-dependent society would be catastrophic. Satellites could be disabled, power grids could collapse, and communication networks could go dark for weeks or even months.

Despite their destructive potential, solar storms also reveal the interconnectedness of the Sun and Earth. The Sun is not merely a distant star; it is a dynamic force that actively shapes the environment of its orbiting planets. The solar wind—a continuous stream of charged particles flowing from the Sun—interacts with Earth's magnetic field, creating the magnetosphere that shields our planet from harmful cosmic radiation. This protective bubble, while essential to life, is also the stage on which solar storms play out.

One of the most remarkable aspects of solar storms is their ability to remind us of our fragility. For all our advancements, we remain deeply vulnerable to the whims of a star. It is humbling to consider that our vast power grids, global communication systems, and technological infrastructure depend on the stability of a Sun that operates on its own terms.

Yet, these storms are not merely destructive. They are also a source of wonder and discovery. By studying solar storms, we gain insights into the fundamental forces of the universe. The phenomena of magnetic reconnection, plasma dynamics, and charged particle acceleration are not unique to the Sun; they occur in stars across the galaxy, in the swirling disks of black holes, and in the luminous jets of quasars. The Sun's storms are a laboratory, offering a glimpse into the universal principles that govern energy and matter.

There is also a philosophical resonance in the storms of the Sun. They remind us of the dual nature of power: its ability to create and to destroy, to illuminate and to obscure. The same energy that allows life to flourish on Earth can, in an instant, disrupt the systems on which we depend.

The Sun

This tension between nurturing and chaos is a fundamental aspect of existence, one that plays out not just on the Sun but in every corner of the universe.

The storms from the Sun also invite us to consider our place in the cosmos. To live in the shadow of such a powerful star is both a privilege and a responsibility. We are shielded by the Earth's magnetic field, protected from the Sun's more destructive impulses, yet we are also exposed to its unpredictable nature. This balance, precarious and delicate, is a reminder of the interconnectedness of all things.

As scientists work to predict and understand solar storms, they are not just safeguarding our technology; they are deepening our understanding of the Sun itself. Each flare, each CME, is a message from the Sun, a glimpse into its inner workings and a reminder of its influence.

The Sun's storms are a paradox. They are born of tension but bring clarity. They are destructive yet beautiful. They remind us of our vulnerability, but also of our capacity for resilience and discovery. To study them is to confront the raw power of the universe—and to find, within that power, the threads of connection that bind us to our star and to each other.

The storms from the Sun are more than bursts of energy. They are a dialogue between a star and its orbiting worlds, a reminder that even in the vastness of space, we are not alone.

The Sun

Chapter 5: The Sun in Myth and Culture

Long before humanity understood the Sun's nuclear heart or magnetic rhythms, we gave it a soul. Across centuries and civilizations, the Sun has been worshipped as a god, feared as a destroyer, and celebrated as the ultimate source of life. Its steady rise and fall have marked the passing of time, while its brilliance has been both a beacon of hope and a symbol of power.

The Sun's place in human culture is not just a reflection of its physical importance but a testament to its symbolic resonance. It is a celestial body that transcends its role as a star to become a mirror for our aspirations, fears, and understanding of the universe.

For ancient peoples, the Sun's daily journey across the sky was a divine act. The Egyptians revered Ra, the Sun god who sailed across the heavens in a golden barge, bringing light and life to the world. In the evening, Ra descended into the underworld, battling chaos to ensure the Sun's return each morning. This myth was not merely a story; it was a way to explain the cosmic order, a reassurance that light would always triumph over darkness.

The Aztecs took their devotion to the Sun to dramatic extremes, performing human sacrifices to ensure the Sun's continued rise. They believed that their Sun god, Huitzilopochtli, required nourishment in the form of human blood to sustain his strength. This belief underscores the profound dependence early civilizations felt toward the Sun, seeing it as both a provider and a force that demanded appeasement.

In ancient Greece, the Sun was personified as Helios, who drove his golden chariot across the sky each day. This imagery of the Sun as a charioteer reflects its perceived agency, its relentless motion, and its vital role in sustaining life. Helios was later supplanted in Greek thought by Apollo, who represented not just the Sun but also art, reason, and prophecy. This duality—linking the Sun with both physical warmth and intellectual illumination—speaks to its centrality in shaping human identity.

The Sun

The Sun's symbolism extends beyond mythology. In many cultures, it has been a marker of kingship and power. The Sun King, Louis XIV of France, adopted the Sun as his emblem, portraying himself as a source of light and life for his realm. His reign embodied the idea of divine right, with the Sun as both his symbol and his justification for rule.

But the Sun has also been a tool for practical innovation. Early agricultural societies used the Sun's cycles to develop calendars, ensuring that planting and harvesting occurred at the optimal times. The Mayans, renowned for their astronomical knowledge, created intricate calendars based on the Sun's movements, aligning their rituals and daily lives with celestial rhythms.

Stonehenge, one of the most famous prehistoric monuments, was constructed with remarkable precision to align with the Sun's solstices. On the summer solstice, the Sun rises in perfect alignment with the monument's central axis, casting light through its ancient stones. This alignment suggests that the builders of Stonehenge understood the Sun's importance not only as a source of warmth and light but also as a guide for marking the passage of time.

Even as scientific understanding grew, the Sun remained a source of artistic and spiritual inspiration. In the Romantic era, poets such as William Blake and Percy Bysshe Shelley used the Sun as a metaphor for divine truth and human creativity. Vincent van Gogh's swirling suns, painted in fiery yellows and oranges, capture its energy and its eternal presence. The Sun, even in its most mundane state, has always been more than a star—it has been a muse.

Today, our relationship with the Sun continues to evolve. We no longer sacrifice to it, but we rely on it just as deeply. Solar energy, harvested through vast arrays of panels, represents humanity's effort to harness the Sun's power in sustainable ways. Meanwhile, solar observatories and spacecraft probe its secrets, transforming what was once divine mystery into scientific discovery.

Yet, for all our advances, the Sun retains an aura of the sacred. A solar eclipse—a phenomenon now understood in precise detail—still draws crowds in awe of its fleeting beauty. The momentary darkening of the sky as the Moon obscures the Sun reminds us of its central role in our

The Sun

lives. The Sun may be a scientific object, but it is also a reminder of nature's power to inspire wonder.

The Sun's cultural resonance lies in its duality. It is both a symbol of life and a reminder of mortality. Its constancy reassures us, but its inevitable transformation into a red giant speaks to the transient nature of all things. These themes resonate deeply across cultures, tying the Sun's story to our own.

In many ways, the Sun is humanity's oldest companion, a constant presence that has shaped not only our physical world but also our spiritual and intellectual landscapes. Its light has guided us, its warmth has sustained us, and its mysteries have challenged us to look beyond ourselves.

To study the Sun in myth and culture is to see the reflection of humanity's journey: from fear to understanding, from reverence to exploration. It is a reminder that even the most familiar objects in the sky can hold infinite layers of meaning. The Sun is not merely a star; it is a symbol of everything we have been and everything we hope to become.

Chapter 6: The Sun's Aging Glow

Stars are not eternal. For all their brilliance and grandeur, they are as mortal as the planets that orbit them, the lifeforms that depend on them, and the civilizations that gaze up at them in wonder. The Sun, too, is finite—a star in its middle age, quietly burning through its supply of hydrogen in a cosmic act of both creation and consumption. Its glow, as steady as it seems, carries the quiet certainty of decline. To understand the Sun's future is to confront the impermanence of all things, including life as we know it.

The Sun is currently in what astronomers call the **main sequence** phase of its life, a stage that accounts for the majority of a star's existence. In this phase, the Sun fuses hydrogen into helium in its core, a process that releases energy and prevents gravitational collapse. This balance between outward pressure from fusion and inward pull from gravity is what keeps the Sun stable.

But this stability is deceptive. Even now, the Sun is changing, though the process is so gradual that it escapes our notice. With every second that passes, it converts approximately 600 million tons of hydrogen into helium. As the helium accumulates in the core, the Sun's fusion processes subtly shift. The core grows denser and hotter, causing the Sun's outer layers to expand and its luminosity to increase. In fact, the Sun is already about 30% brighter than it was when it first formed, a change that has had profound implications for Earth's climate and biosphere over geological time.

Fast-forward to the distant future, and the changes become impossible to ignore. In about a billion years, the Sun's increasing brightness will render Earth uninhabitable. The oceans will boil away, the atmosphere will be stripped, and the surface will become a scorched wasteland. Life, if it still exists, will have to adapt or migrate elsewhere. By this time, humanity may no longer be tethered to Earth, but the Sun's evolution will continue regardless of our presence.

The Sun's middle age will eventually give way to its red giant phase, a transformation that is both spectacular and apocalyptic. When the Sun exhausts the hydrogen in its core, fusion will cease in that region, and

The Sun

gravity will begin to compress the core. This compression will heat the surrounding hydrogen shell, igniting fusion in a new layer around the inert helium core. The energy released from this shell fusion will cause the Sun's outer layers to expand dramatically.

During its red giant phase, the Sun will swell to hundreds of times its current size, engulfing Mercury and Venus—and possibly Earth. The solar wind, a stream of charged particles emitted by the Sun, will intensify, stripping away the atmospheres of any surviving inner planets. Mars, if it remains outside the Sun's reach, may experience a brief renaissance, its icy surface thawing under the Sun's bloated warmth.

The Sun's red giant phase, though catastrophic for the inner solar system, is not its final act. When the helium in the core ignites, it will produce carbon and oxygen in a process known as helium fusion. But this phase will be short-lived, as the Sun lacks the mass to sustain further fusion processes. Once the helium is depleted, the Sun will shed its outer layers in a series of dramatic pulses, creating a planetary nebula—a glowing shell of gas that will briefly illuminate the surrounding cosmos.

What remains will be a white dwarf, a dense, Earth-sized remnant composed primarily of carbon and oxygen. This white dwarf, no longer capable of fusion, will slowly cool and fade over billions of years, becoming a black dwarf—a cold, dark cinder in the vastness of space. The journey from brilliance to darkness will have taken billions of years, but it will end in silence.

The Sun's aging glow is not just a story of cosmic mechanics; it is a profound metaphor for the cycles of creation and decay that define the universe. It is a reminder that even the most enduring forces are subject to change, that even stars must face their mortality.

This narrative also challenges our sense of time and permanence. The Sun's lifespan, vast as it seems, is but a blink in the history of the cosmos. Stars far older than the Sun have already lived and died, their remnants seeding the universe with the elements that make up planets, oceans, and life itself. The Sun, in its turn, will contribute to this cosmic cycle, dispersing its material to form new stars and worlds.

The Sun

For humanity, the Sun's eventual demise raises questions that go beyond science. Will we, as a species, endure long enough to witness these changes? Will we migrate to other stars, carrying the memory of our solar origin? Or will we fade alongside our Sun, a brief flicker in the grand tapestry of existence?

There is a certain beauty in the Sun's inevitable decline. It is not a failure but a natural evolution, a transformation that echoes across the cosmos. The death of a star is not an end; it is a beginning, the catalyst for new creations in the vastness of space. The Sun's legacy will live on in the light of new stars, the formation of new planets, and the possibility of life arising elsewhere.

In the Sun's aging glow, we see both the fragility and the resilience of the universe. It is a story that humbles us, reminding us of our place in the grand scheme of things, and a story that inspires us, showing us the interconnectedness of all existence. The Sun, even in its final stages, will remain a beacon—a symbol of the cycles that shape not just stars, but life itself.

Chapter 7: The Science of Solar Observation

For millennia, the Sun was an object of reverence and mystery, its power revered in temples and its light used to chart the heavens. But it was not until humanity developed the tools of science that we began to truly see the Sun—not as a god, but as a star. The science of solar observation has transformed the Sun from an untouchable presence into a subject of study, revealing its intricate workings and profound influence on the universe. Yet, even as we uncover its secrets, the Sun continues to elude full understanding, challenging us with mysteries that push the boundaries of science and imagination.

The journey to understanding the Sun began with our earliest attempts to track its movements. Ancient cultures built observatories—such as Stonehenge in England and the pyramids of Egypt—designed to mark the solstices and equinoxes. These structures were not merely architectural marvels; they were tools for understanding the Sun's rhythms and integrating them into daily life. In these early observations, humanity began to realize that the Sun was not a capricious force but a predictable one, its cycles woven into the fabric of existence.

The invention of the telescope in the 17th century was a turning point. With this simple but revolutionary instrument, Galileo Galilei became one of the first to observe sunspots—dark blemishes on the Sun's surface. His meticulous recordings of their movements revealed that the Sun rotated on its axis, a discovery that challenged long-held assumptions about the heavens. Galileo's work, controversial in his time, laid the foundation for a new era of solar science, one that sought to understand the Sun not as a divine entity but as a celestial object governed by natural laws.

The 19th and 20th centuries brought new tools and techniques that transformed solar observation. Spectroscopy allowed scientists to analyze the Sun's light and determine its composition, revealing that it is primarily made of hydrogen and helium. This discovery, combined with advances in physics, led to the groundbreaking realization that the Sun's

The Sun

energy comes from nuclear fusion. For the first time, humanity understood the engine that powers the Sun—a process so powerful it defies imagination.

Yet, even with these advancements, observing the Sun remained a formidable challenge. Its brilliance, so vital to life, is also a barrier to study. Looking directly at the Sun without protection can damage the eyes, and the sheer intensity of its light obscures the finer details of its structure. To overcome these obstacles, scientists developed specialized instruments, such as solar filters and coronagraphs, which block the Sun's overwhelming light to reveal its outer layers and corona.

In recent decades, spacecraft have brought us closer to the Sun than ever before. NASA's Parker Solar Probe, launched in 2018, is currently orbiting the Sun, gathering data from its outer atmosphere. It is the closest any human-made object has ever come to our star, braving temperatures of over 1,000 degrees Celsius to unlock its secrets. Similarly, the Solar and Heliospheric Observatory (SOHO), a joint mission between NASA and the European Space Agency, has provided a continuous stream of data about the Sun's activity since 1995, revolutionizing our understanding of solar phenomena.

These missions have answered some questions but raised others. Why is the corona, the Sun's outer atmosphere, so much hotter than its surface? What drives the Sun's magnetic cycles, and why do they vary in intensity? How do solar storms affect planets beyond Earth? Each answer seems to uncover new layers of complexity, reminding us that the Sun, for all its proximity, remains a frontier of discovery.

The science of solar observation is not just about understanding the Sun; it is about understanding the universe. The processes that govern the Sun are not unique to our star; they are fundamental to stars across the galaxy. By studying the Sun, we gain insights into the formation and evolution of other stars, the dynamics of stellar systems, and the origins of the elements that make up planets and life itself.

Solar observation also connects us to the practical realities of living in the Sun's domain. Space weather, driven by solar activity, affects everything from satellite communications to power grids, posing both challenges and opportunities. By predicting solar storms, we can protect

The Sun

our technology-dependent world, ensuring that the Sun's occasional fury does not catch us unprepared.

But there is another dimension to solar observation—one that transcends science and touches the realm of philosophy. To gaze at the Sun is to confront a force that is both familiar and alien. It is a reminder of our dependence on a star that is, in many ways, indifferent to our existence. The act of observing the Sun is a dialogue with the cosmos, an attempt to understand not just the star itself but our place within its vast sphere of influence.

There is also a profound beauty in the Sun's complexity. The granules on its surface, the arcs of plasma in its corona, the vibrations that ripple through its core—these are not just phenomena to be measured but expressions of a dynamic universe. The Sun is not a static object but a living system, its movements and rhythms echoing the cycles of creation and decay that shape all things.

The science of solar observation is a testament to human curiosity and ingenuity. It is the story of how a species, bound to a small blue planet, reached out to understand the star that gave it life. It is a story of light and shadow, of questions answered and mysteries revealed. And it is a reminder that even in the face of something as immense as the Sun, the act of looking, questioning, and discovering is an act of profound significance.

As we continue to study the Sun, we are not merely observing a star. We are exploring the forces that shape existence itself, finding in its brilliance the threads that connect us to the universe. The Sun, in its endless glow, invites us to see not just what it is, but what it can teach us about everything.

Chapter 8: The Sun's Dual Nature

The Sun is a paradox. It is the giver of life, yet it holds the power to destroy. It is a symbol of constancy, rising and setting without fail, yet its surface is a cauldron of chaos and unpredictable violence. For billions of years, the Sun has sustained Earth, providing light, warmth, and energy. But this same star, so essential to life, also threatens it with storms, radiation, and an inevitable death that will unmake everything it has nurtured. To understand the Sun is to grapple with this duality, to accept that it embodies both creation and destruction, harmony and discord.

The Sun's life-giving role is obvious. Without its light, Earth would be a frozen wasteland, incapable of supporting the diverse ecosystems that flourish across its surface. Photosynthesis, the process by which plants convert sunlight into energy, forms the foundation of nearly every food chain. The Sun's warmth drives weather patterns, circulates ocean currents, and sustains the delicate balance that makes Earth habitable.

Yet, even this apparent harmony is precarious. The Sun is not a gentle nurturer; it is an indifferent force, bound by the laws of physics rather than any sense of purpose. Solar storms, born of magnetic turmoil, remind us of the Sun's potential for disruption. These storms are not malicious—they are simply the Sun being itself—but their effects on Earth can be profound, disrupting communication systems, power grids, and navigation technologies.

This duality becomes even more apparent when we consider the Sun's distant future. While it currently resides in a stable middle age, the Sun is steadily growing brighter. In a billion years, its increased luminosity will render Earth uninhabitable, boiling away the oceans and stripping the atmosphere. Eventually, the Sun will swell into a red giant, consuming Mercury and Venus—and perhaps Earth—in its fiery embrace. What began as a life-giving star will become an agent of destruction, erasing the very worlds it once illuminated.

This transformation is not unique to the Sun. It is the fate of all stars like it, a reminder that creation and destruction are not opposites but two sides of the same cosmic coin. The elements that make up planets, oceans,

The Sun

and living organisms—carbon, oxygen, nitrogen—were forged in the hearts of ancient stars that lived, burned, and died long before the Sun was born. These elements were scattered into space during supernovae, seeding the raw materials for new stars and worlds. The Sun itself will one day contribute to this cycle, its outer layers cast off into space to form a nebula, while its core collapses into a dense white dwarf.

This interplay of birth and death, growth and decay, is not limited to the Sun. It is a universal principle, echoed in the cycles of life on Earth. Plants grow, flourish, and wither, returning their nutrients to the soil. Stars ignite, burn, and fade, giving rise to new generations of celestial bodies. Even civilizations rise and fall, leaving behind the seeds of future cultures. The Sun's dual nature is a reflection of this cosmic truth: that destruction is not an end, but a transformation.

There is a profound lesson in this duality. The Sun does not choose to give or take; it simply exists, following the laws of nature. Its light sustains life not out of benevolence, but because of its inherent properties. Its storms are not punishments, but consequences of its dynamic behavior. This indifference is humbling, reminding us that the universe is vast and impersonal. Yet, it is precisely this indifference that makes life so extraordinary.

Humanity's relationship with the Sun embodies this duality. We depend on its energy, yet we strive to shield ourselves from its extremes. Solar panels harness its light to power our homes and cities, while sunscreen protects us from its harmful ultraviolet rays. Spacecraft and observatories monitor the Sun's activity, allowing us to predict solar storms and mitigate their impact. In doing so, we are both embracing and resisting the Sun's influence, seeking to live in balance with a force that is neither friend nor foe.

This balance is precarious, yet it also offers a unique perspective. The Sun, for all its power, is finite. Its energy, though immense, is drawn from a limited supply of hydrogen that will one day run out. In this, the Sun is not so different from us. It, too, has a lifespan, a beginning and an end, a story shaped by the interplay of creation and destruction.

The Sun's dual nature also invites us to reflect on our own. Humanity, like the Sun, is capable of great creation and profound destruction. We

The Sun

build civilizations, invent technologies, and explore the universe, yet we also exploit resources, wage wars, and alter ecosystems. In the Sun, we see a mirror of our own complexity—a reminder that our greatest strengths are often tied to our greatest vulnerabilities.

But the Sun's duality is not something to fear. It is something to understand, to marvel at, and to learn from. Its cycles, its storms, and its eventual transformation are not acts of malice or benevolence; they are expressions of the fundamental laws of existence. In accepting this, we find a deeper appreciation for the Sun—not just as a star, but as a teacher.

The Sun's dual nature is a story of balance and change, constancy and evolution. It is a reminder that life, in all its fragility, exists within a universe that is both nurturing and indifferent. To study the Sun is to see this balance in action, to witness the intricate dance of forces that shape our world and our place within it.

The Sun gives, and the Sun takes. It sustains, and it destroys. And in its light and shadow, we find the truth of our own existence—a truth that is at once humbling and inspiring.

Chapter 9: The Sun's Dual Role in the Cosmos

The Sun is more than the center of our solar system. It is a cosmic force, a star among billions in the Milky Way, whose existence echoes through the universe. Its influence extends far beyond the light that graces our skies each day. The Sun is both a creator and a guardian, a force that shapes not just the planets in its orbit but the wider structure of the galaxy. Yet, it is also a star among many, one whose significance depends entirely on perspective.

In the immediate sense, the Sun is the architect of our solar system. Some 4.6 billion years ago, its birth set the stage for the formation of planets, moons, and asteroids. As gravity pulled gas and dust into the nascent star, the remaining material coalesced into the celestial bodies that now share the Sun's orbit. The Sun's energy ignited life on Earth and governs the cycles of every world in its domain. Without it, the solar system would be a cold, dark void—a collection of barren rocks drifting aimlessly through space.

The Sun's gravitational pull binds the solar system together, maintaining the orbits of its planets and keeping them on their steady paths. It is not an exaggeration to say that the Sun is the solar system's anchor. Without its stabilizing influence, the delicate balance of planetary motions would unravel, leading to chaotic collisions and the loss of worlds.

But the Sun's role as a guardian goes beyond gravity. Its heliosphere, a vast bubble of solar wind and magnetic fields, acts as a protective shield for the planets. This heliosphere deflects cosmic rays—high-energy particles originating from outside the solar system—that could otherwise wreak havoc on planetary atmospheres and living organisms. In this sense, the Sun is not just a source of energy but also a protector, a barrier against the hostile forces of interstellar space.

The Sun's dual role as creator and guardian becomes even more intriguing when we consider its relationship with Earth. While the Sun's energy is essential for life, its variability presents a challenge. Solar storms, caused by eruptions of plasma and magnetic fields from the Sun's

The Sun

surface, can disrupt communication systems, power grids, and even human health. These storms are reminders that the Sun, for all its nurturing qualities, is not entirely benign.

This duality extends to the Sun's future. As it evolves, it will continue to shape the solar system in profound ways. When the Sun eventually swells into a red giant, it will obliterate some of its closest planets while potentially giving new life to others. Mars, currently a cold and arid world, may temporarily find itself in the habitable zone, its surface thawing under the Sun's expanded warmth.

The Sun's eventual death will scatter its outer layers into space, forming a planetary nebula that will enrich the interstellar medium with elements forged in its core. This material will become the building blocks of new stars, planets, and possibly life. In this way, the Sun's end will mark a new beginning, its legacy written in the stars that follow.

The Sun's role in the galaxy, however, is not unique. It is one of billions of stars in the Milky Way, each playing its part in the intricate structure of the cosmos. The Sun's position—nestled in a relatively quiet region of the galactic disk—is one of the factors that has allowed life to thrive on Earth. In more chaotic regions, such as the galactic core, the density of stars and the prevalence of high-energy radiation would make life as we know it impossible.

This perspective raises a humbling question: Is the Sun special? From the vantage point of Earth, the answer seems obvious—it is the star that sustains us, the center of our world. But in the grand scale of the universe, the Sun is an ordinary star, a G-type main-sequence star like countless others. Its light is unremarkable, its lifespan typical. It is one among many, its significance defined not by its intrinsic qualities but by its relationship with the worlds that orbit it.

This duality—between the Sun's cosmic insignificance and its local centrality—reflects a broader truth about existence. Meaning is not an inherent quality but a relationship. The Sun matters to us because we depend on it, because we are tied to its rhythms and its light. Yet, in the vastness of the universe, the Sun is no more important than any other star.

The Sun

There is beauty in this perspective. The Sun, ordinary as it may be in the galactic sense, is extraordinary to us. It is a reminder that significance is not found in absolute terms but in connection. The Sun is our star, the one that gave rise to our planet and our lives. Its ordinary nature does not diminish its importance; it amplifies it, showing that even the most common forces can create extraordinary things.

The Sun's dual role—as a local creator and a galactic participant—is a testament to the interconnectedness of the universe. It reminds us that we are part of something far greater than ourselves, a web of relationships that spans light-years and eons. The Sun's light, traveling across millions of kilometers to reach Earth, is a bridge between the cosmic and the intimate, the infinite and the immediate.

In understanding the Sun's dual role, we begin to see our own. Just as the Sun is both ordinary and extraordinary, so too are we. Our lives are shaped by forces larger than ourselves, yet we have the power to shape meaning within our small corner of the cosmos. The Sun, in its light and shadow, its creation and destruction, is a mirror for our own complexities—a reminder that we, too, are part of the universe's story.

The Sun's role in the cosmos is not fixed. It evolves, it transforms, it leaves a legacy that will outlast its glow. And in its light, we find the threads that connect us to the stars, to the galaxy, and to the vast expanse of existence itself.

Conclusion – A Star's Legacy

The Sun is both timeless and transient, a force that has shaped the solar system for billions of years and will continue to do so long after humanity's story is written. It is more than a star—it is a presence that anchors not just planets but the human experience, infusing every corner of life with its light and energy. Yet, for all its brilliance, the Sun's legacy lies not in its unchanging nature, but in its evolution. It is a star in motion, a dynamic entity whose story is as much about change as it is about constancy.

Through the chapters of this book, we have traced the Sun's journey from its fiery birth in a collapsing cloud of gas to its current state as a middle-aged star, quietly burning through its nuclear fuel. We have marveled at its magnetic cycles, explored its chaotic storms, and glimpsed the intricate web of its influence that extends from the Earth's surface to the farthest reaches of the heliosphere. Each revelation brings us closer to understanding not just the Sun, but the forces that govern the universe itself.

Yet, the Sun's story is far from over. Its eventual transformation into a red giant will mark a dramatic chapter in the solar system's history, reshaping planets and scattering its material into the interstellar medium. This is not the end, but a continuation—a passing of the torch, as the Sun's remnants contribute to the formation of new stars, new planets, and perhaps new life.

The Sun's legacy is written in cycles, each one echoing the rhythms of the cosmos. The 11-year solar cycle, with its flares and magnetic reversals, reflects the tension between order and chaos that defines the Sun's nature. Its lifecycle, from main sequence to red giant to white dwarf, mirrors the universal pattern of birth, growth, decay, and renewal. These cycles remind us that nothing is permanent, but everything is connected.

The Sun's light, the most familiar aspect of its existence, also carries its legacy forward. Traveling at the speed of light, it takes just over eight minutes for photons to reach Earth, yet the journey of these particles began thousands of years earlier, deep within the Sun's core. Long after

The Sun

the Sun has burned out, the light it emitted will continue to travel through the universe, carrying with it the story of our star.

For humanity, the Sun is more than a celestial object. It is a symbol of life and a source of wonder, a force that has inspired myths, driven scientific discovery, and shaped the rhythms of civilization. From the agricultural calendars of ancient cultures to the solar panels of the modern age, the Sun has been a constant companion in our quest to understand the world and our place within it.

But the Sun is also a humbling presence. Its vastness dwarfs our planet, its energy outstrips anything humanity can imagine, and its indifference reminds us of the fragility of life. To live in the Sun's light is both a privilege and a challenge, an existence balanced between dependence and vulnerability.

The legacy of the Sun, however, is not just its own. It is also ours. The story of the Sun is the story of Earth, of life, and of humanity's endless curiosity. By studying the Sun, we uncover not just its secrets but our own, finding in its light the threads that connect us to the universe. The Sun teaches us that we are part of something vast and dynamic, a cosmos where creation and destruction are inseparable, and where meaning arises from connection.

In its brilliance and its shadow, the Sun is a reminder of the delicate balance that sustains life and the cycles that define existence. Its legacy is not just its light, but the lessons it offers: about resilience, impermanence, and the profound beauty of the universe.

As we look to the Sun, we see not just a star, but a story—one that began long before us and will continue long after we are gone. It is a story of light and shadow, of constancy and change, and of the enduring connection between a star and the worlds it illuminates.

The Sun's legacy is a cosmic reminder that even in the vastness of space, there is meaning, there is purpose, and there is life.

End Note: Beyond the Sun

The Sun is our beginning, but it is not our end. As humanity peers deeper into the cosmos, we discover that the story of our star is echoed in countless others. The Sun, though unique to us, is but one actor in an infinite drama, a single voice in the chorus of stars that fill the universe. Its light, its energy, its cycles—these are not just phenomena confined to our solar system, but threads in the vast tapestry of existence.

In studying the Sun, we learn more than the mechanics of a star; we learn about the forces that shape the cosmos itself. The fusion processes that power the Sun illuminate the origins of the elements, the same carbon, oxygen, and nitrogen that compose our bodies and our world. The magnetic storms that ripple across its surface reveal universal principles of energy and matter, echoed in the distant galaxies and nebulae that inspire our curiosity.

But the Sun also teaches us something deeper. It reminds us that life is a precarious balance, sustained by forces we can barely control. It reminds us of our resilience, our capacity to adapt and thrive in the light of a star that is both nurturing and indifferent. And it reminds us of the interconnectedness of all things—that the energy that fuels a blade of grass on Earth has its origin in the heart of a star, that the atoms in our bodies were forged in stellar furnaces, and that our destiny, like the Sun's, is tied to the cycles of the universe.

As we gaze at the Sun, we are reminded of our dual nature. We are creatures of Earth, bound to this world by gravity and biology, yet we are also creatures of the stars, our curiosity pulling us outward, our imagination reaching for the infinite. The Sun is both a reminder of our origins and a guide to our future, illuminating the path to understanding not just the cosmos, but ourselves.

In its light, we find the answers to some of our most profound questions, and in its shadow, we find the mystery that drives us to seek more. The Sun is not just a star; it is a symbol of the endless interplay between the known and the unknown, a reflection of the cycles that define all existence.

The Sun

As this book comes to a close, the journey it represents continues. The Sun's light will keep traveling, its energy will keep creating, and its story will keep unfolding, long after we are gone. And in that story, we find not just the legacy of a star, but the promise of discovery—a promise that, as long as we look to the sky, the light of the Sun will always guide us.

www.ingramcontent.com/pod-product-compliance
Lightning Source LLC
Chambersburg PA
CBHW070945220526
45469CB00007B/2521